NOT ROCKET SCIENCE

Prof. ODDFellow's
Forgotten Wisdom

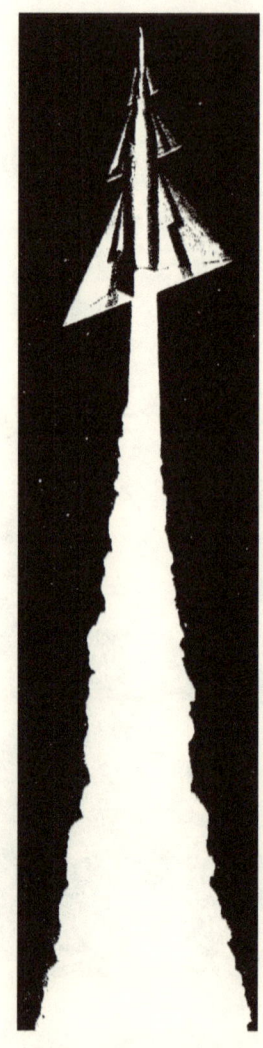

Inspired by
&~ dedicated to
Martha Brockenbrough

PROVING WOMEN ALSO HAVE IDEAS

"It was underwear,
not rocket science.
Or was it?"

—Jennifer Barnes,
The Squad, 2008

ROCKET SCIENCE?

Yes	No	Maybe

☺ ☹ 😐 **Looking for a new job.**

—Tod Bermont, *10 Insider Secrets to a Winning Job Search*, 2004, p. 47

☺ ☹ 😐 **Sending out a résumé.**

—Susan Ireland, *The Complete Idiot's Guide to the Perfect Resume*, 2003, p. 176

☺ ☹ 😐 **Making tortillas.**

—Carl Franz, *The People's Guide to Mexico*, 2006, p. 147

☺ ☹ 😐 **Providing Ritz-Carlton service with FedEx efficiency.**

—Matt Oechsli, *The Art of Selling to the Affluent*, 2005, p. 146

ROCKET SCIENCE?

Yes	No	Maybe

☺ ☹ 😐   **Booking travel.**

—Peter Greenberg, *The Complete Travel Detective Bible*, 2007, p. 206

☺ ☹ 😐   **Taking care of yourself in the woods.**

—Glenn Scherer, *Hikes in the Mid-Atlantic States, 1998*, p. 23

☺ ☹ 😐   **Who gets what, when, where, why, and how much.**

—Robert Bullard, *Highway Robbery*, 2004, p. 5

ROCKET SCIENCE?

Yes   No   Maybe

☺   ☹   😐   **Basic principles of rocketry.**

—A. Bowdoin Van Riper, *Rockets and Missiles*, 2004, p. 2

☺   ☹   😐   **Licensing a million-dollar idea.**

—Harvey Reese, *How to License Your Million Dollar Idea*, 2002, p. 32

☺   ☹   😐   **The race for space.**

—Betsy Kuhn, *The Race for Space*, 2007, p. 24

☺   ☹   😐   **Launching, flying and returning the Space Shuttle.**

—Fergus O'Connell, *Simply Brilliant*, 2004, p. 5

ROCKET SCIENCE?

Yes	No	Maybe
☺	☹	😐

The solution to hunger.

—Sharman Russell, *Hunger*, 2005, p. 214

Yes	No	Maybe
☺	☹	😐

Eating.

—Brooke Castillo, *If I'm So Smart, Why Can't I Lose Weight?*, 2006, p. 15

Yes	No	Maybe
☺	☹	😐

Stir-frying.

—Martin Yan, *Martin Yan Quick and Easy*, 2004, p. 14

Yes	No	Maybe
☺	☹	😐

Slow cooker cooking.

—Ellen Brown, *The Complete Idiot's Guide to Slow Cooker Cooking*, 2003, p. 4

Yes	No	Maybe
☺	☹	😐

Homebrewing.

—Marty Nache, *Homebrewing for Dummies*, 2008, p. 59

Yes	No	Maybe
☺	☹	😐

Dieting.

—Donna Rodnitzky, *The Ultimate Low-Carb Diet Cookbook*, 2001, p. 3

ROCKET SCIENCE?

Yes   No   Maybe

☺   ☹   😐   **Enjoying yourself at a party.**

—Michael Flocker, *The Metrosexual Guide to Style*, 2003, p. 5

☺   ☹   😐   **Booze (and bartending).**

—Eileen Rendahl, *Un-Bridaled*, 2006, p. 108

☺   ☹   😐   **Marriage proposals.**

—Phineas Mollod, *The Modern Lover*, 2004, p. 180

ROCKET SCIENCE?

Yes	No	Maybe	
☺	☹	😐	**Identifying someone who is jobless.**

—Bernard Baumohl, *The Secrets of Economic Indicators*, 2007, p. 27

☺	☹	😐	**Getting a date worth keeping.**

—Henry Cloud, *How to Get a Date Worth Keeping*, 2005, p. 225

☺	☹	😐	**Lodging a complaint about another's failure to render service.**

—Martin Kevorkian, *Color Monitors*, 2006, p. 76

☺	☹	😐	**Learning like a girl.**

—Diana Meehan, *Learning Like a Girl*, 2003, p. 88

☺	☹	😐	**Success with women.**

—Ron Louis, *How to Succeed with Women*, 1998, p. 18

ROCKET SCIENCE?

Yes   No   Maybe

☺   ☹   😐   **Getting things done.**

—David Allen, *Getting Things Done*, 2003, p. 86

☺   ☹   😐   **Winning the race against time.**

—Dan Carison, *Deadline!*, 2002, p. 96

ROCKET SCIENCE?

Yes	No	Maybe
☺	☹	😐

Driving.

—Janet Chapman, *The Seduction of His Wife*, 2006, p. 151

☺	☹	😐

Being a champion car racer.

—Tony Stewart, *High Octane in the Fast Lane*, 2002, p. 106

☺	☹	😐

Nose jobs.

—Marissa Monteilh, *Make Me Hot*, 2008, p. 71

ROCKET SCIENCE?

Yes   No   Maybe

☺   ☹   😐   **Firing a gun.**

—John Hart, *The King of Lies*, 2007, p. 256

☺   ☹   😐   **Shooting a soup can at fifty yards.**

—Eric Flint, *Ring of Fire*, 2003

☺   ☹   😐   **Counterinsurgency warfare.**

—David Galula, *Counterinsurgency Warfare*, 2006, p. viii

ROCKET SCIENCE?

Yes   No   Maybe

☺   ☹   😐   **Negotiation.**

—Steve Cohen, *Negotiating Skills for Managers*, 2002, p. 168

☺   ☹   😐   **Outsourcing.**

—Bob Booton, *Outsourcing-in-a-Box*, 2005, p. 97

☺   ☹   😐   **Investing.**

—Stacey Tisdale, *The True Cost of Happiness*, 2007, p. 199

☺   ☹   😐   **Wealth-building habits.**

—Jason Trennert, *New Markets, New Strategies*, 2004, p. 132

ROCKET SCIENCE?

Yes   No   Maybe

☺ ☹ 😐 **Branding.**

—Kevin Keller, *Strategic Brand Management*, 1998, p. xv

☺ ☹ 😐 **Retail pricing.**

—David Evans, *Paying with Plastic,* 2005, p. 139

☺ ☹ 😐 **Selling jeans.**

—Paul Trott, *Innovation Management and New Product Design*, 2005, p. 144

☺ ☹ 😐 **Adolescents' sexual decisions.**

—Mark Regnerus, *Forbidden Fruit*, 2007, p. 84

☺ ☹ 😐 **Adams College admissions.**

—Burton Nadler, *The Adams College Admissions Essay Handbook*, 2004, p. 1

ROCKET SCIENCE?

Yes	No	Maybe
☺	☹	😐

Genetics.
—Ingrid Wood, *A Breeder's Guide to Genetics*, 2004

Yes	No	Maybe
☺	☹	😐

Neonatology.
—Mark Davies, *Pocket Notes on Neonatology*, 2008, p. 30

Yes	No	Maybe
☺	☹	😐

Child-focused work.
—Judith Palfrey, *Child Health in America*, 2006, p. 221

Yes	No	Maybe
☺	☹	😐

Adopting from Russia, Ukraine, and Kazakhstan.
—John MacLean, *The Russian Adoption Handbook*, 2004, p. 97

Yes	No	Maybe
☺	☹	😐

Nannying.
—Haven Kimmel, *The Solace of Leaving Early*, 2003, p. 59

ROCKET SCIENCE?

Yes   No   Maybe

☺   ☹   😐   **Men's health.**

—Lloyd Bradley, *Not Rocket Science: Men's Health*, 2005

☺   ☹   😐   **Health care reform.**

—David Mechanic, *The Truth About Health Care*, 2006, p. 188

☺   ☹   😐   **Magnetic Resonance.**

—Donald McRobbie, *MRI from Picture to Proton*, 2007, p. 411

☺   ☹   😐   **Looking after end-stage heart failure.**

—Miriam Johnson, *Heart Failure and Palliative Care*, 2006, p. vi

ROCKET SCIENCE?

Yes   No   Maybe

☺   ☹   😐   **Occupying Iraq.**
—James Loewen, *Lies My Teacher Told Me*, 2007. p. 276

☺   ☹   😐   **Plan-as-you-go business planning.**
—Tim Berry, *The Plan-as-You-Go Business Plan*, 2008, p. 178

☺   ☹   😐   **Risk assessment.**
—Richard Claude, *Science in the Service of Human Rights*, 2002, p. 141

☺   ☹   😐   **Fire investigation.**
—Shannon Cowan, *Tin Angel*, 2007, p. 299

☺   ☹   😐   **Media relations.**
—David Henderson, *Making News*, 2006, p. 190

ROCKET SCIENCE?

Yes No Maybe

☺ ☹ 😐 **Acceptable moral values.**
—Jon Huntsman, *Winners Never Cheat*, 2005, p. 173

☺ ☹ 😐 **Studying the Greek manuscripts of the New Testament.**
—Bart Ehrman, *Misquoting Jesus*, 2005, p. 77

☺ ☹ 😐 **Pointing people to faith.**
—Bill Hybels, *Just Walk Across the Room*, 2006, p. 105

☺ ☹ 😐 **Coexisting.**
—Jim Merkel, *Radical Simplicity*, 2003, p. 58

☺ ☹ 😐 **Practical paganism.**
—Dana Eilers, *The Practical Pagan*, 2002, p. 38

ROCKET SCIENCE?

Yes   No   Maybe

☺ ☹ 😐 **Ecology.**

—William Dritschilo, *Earth Days*, 2004, p. 79

☺ ☹ 😐 **Clean water.**

—Oliver Houck, *The Clean Water Act TMDL Program*, 2002, p. 143

☺ ☹ 😐 **The study of soils.**

—Shane Smith, *Greenhouse Gardener's Companion*, 2000, p. 353

☺ ☹ 😐 **Forest management.**

—David Lindemayer, *Practical Conservation Biology*, 2005, p. 6

☺ ☹ 😐 **Insect rearing.**

—Allen Cohen, *Insect Diets*, 2004, p. 14

☺ ☹ 😐 **Gathering eggs.**

—Anna Jeffrey, *Sweet Return*, 2007, p. 141

☺ ☹ 😐 **Building a sustainable business.**

—Mitchell Goozé, *It's Not Rocket Science*, 1997

ROCKET SCIENCE?

Yes	No	Maybe	
☺	☹	😐	**Archaeology.**

—Larry Zimmerman, *Ethical Issues in Archaeology*, 2003, p. 216

☺	☹	😐	**Statutory interpretation.**

—Guyora Binder, *Literary Criticisms of Law*, 2000, p. 81

☺	☹	😐	**Reconstructing contexts.**

—Robert Hume, *Reconstructing Contexts*, 1999, p. 99

☺	☹	😐	**Reverse-engineering.**

—Jessica Livingston, *Founders at Work*, 2008, p. 3

ROCKET SCIENCE?

Yes   No   Maybe

☺   ☹   😐   **Learning to read.**

—Frank Smith, *Understanding Reading*, 2004, p. 3

☺   ☹   😐   **Elizabethan literature.**

—Michele Albert, *Hide in Plain Sight*, 2006, p. 39

ROCKET SCIENCE?

Yes No Maybe

🙂 🙁 😐 **Designing a 360-degree questionnaire.**
—Clive Fletcher, *Appraisal and Feedback*, 2004, p. 69

🙂 🙁 😐 **Asking open-ended questions and letting someone talk.**
—Jerry Acuff, *The Relationship Edge in Business*, 2004, p. 78

🙂 🙁 😐 **Listening, understanding, and validating.**
—Greg Smalley, *The Marriage You've Always Dreamed Of*, 2005, p. 167

🙂 🙁 😐 **Linguistics.**
—John McWhorter, *Losing the Race*, 2000, p. 141

ROCKET SCIENCE?

Yes	No	Maybe	
☺	☹	😐	**Running.**

—Jack Nirenstein, *Running*, 2005, p. 16

| ☺ | ☹ | 😐 | **Pacesetter performance.** |

—Rex Kenyon, *Process Plant Reliability and Maintenance for Pacesetter Performance*, p. xvi

| ☺ | ☹ | 😐 | **Diagnosing chronic pain.** |

—Arthur Rosenfeld, *The Truth About Chronic Pain*, 2003, p. 147

ROCKET SCIENCE?

Yes	No	Maybe
☺	☹	😐

Playing the violin.

—Kevin Leman, *Sheet Music*, 2003, p. 63

| ☺ | ☹ | 😐 |

Booking your own concerts.

—Angela Beeching, *Beyond Talent*, 2005, p. 147

| ☺ | ☹ | 😐 |

Producing great sound for digital video.

—Jay Rose, *Producing Great Sound for Digital Video*, 2002, p. xi

| ☺ | ☹ | 😐 |

Editing video.

—Cody McFadyen, *Shadow Man*, 2007, p. 49

| ☺ | ☹ | 😐 |

Guerilla marketing.

—Jay Levinson, *Guerilla Marketing in 30 Days*, 2005, p. 159

ROCKET SCIENCE?

Yes	No	Maybe	
☺	☹	😐	**Quality improvement.** —Robert Cole, *Managing Quality Fads*, 1999, p. 82
☺	☹	😐	**Off-balance-sheet financing.** —H. David Sherman, *Profits You Can Trust*, 2003, p. 71
☺	☹	😐	**Keeping your staff.** —Conrad Lashley, *Business Development in Licensed Retailing*, 2003, p. 108

ROCKET SCIENCE?

Yes No Maybe

☺ ☹ 😐 **Relationship building.**

—Debbie Bermont, *Outrageous Business Growth*, 2004, p. 181

☺ ☹ 😐 **Doing laundry.**

—Wendy Walker, *Four Wives*, 2008, p. 115

☺ ☹ 😐 **Cleaning up.**

—Eve Salinger, *The Complete Idiot's Guide to Pleasing Your Woman*, 2006, p. 102

ROCKET SCIENCE?

Yes   No   Maybe

☺   ☹   😐   **Low-slope roof systems.**

—Charles Griffin, *Manual of Low-Slope Roof Systems*, 2006, p. xiii

☺   ☹   😐   **Septic system design.**

—Lloyd Kahn, *The Septic System Owner's Manual*, 2007, p. 61

☺   ☹   😐   **Hazard and security planning.**

—Transportation Research Board, *Public Transportation Security*, 2006

ROCKET SCIENCE?

Yes   No   Maybe

☺   ☹   😐   **Education.**

—D. B. Zandvliet, *Education is Not Rocket Science*, 2006

☺   ☹   😐   **Authentic pedagogy facing the Other of its own identity.**

—David Smith, *Trying to Teach in a Season of Great Untruth*, 2006, p. 71

ROCKET SCIENCE?

Yes   No   Maybe

☺   ☹   😐   **Running a major college basketball program.**

—Jim Calhoun, *A Passion to Lead*, 2007, p. 36

☺   ☹   😐   **Coaching across cultures.**

—Philippe Rosinski, *Coaching Across Cultures*, 2003, p. 261

ROCKET SCIENCE?

Yes   No   Maybe

☺   ☹   😐   **Dog training.**

—Gerilyn Bielakiewicz, *The Everything Dog Training and Tricks Book*, 2002, p. 178

☺   ☹   😐   **Cat care.**

—Elizabeth Hodgkins, *Your Cat*, 2007, p. 147

☺   ☹   😐   **Herding chickens.**

—David Garrett, *Herding Chickens*, 2005, p. 128

ROCKET SCIENCE?

Yes   No   Maybe

☺   ☹   😐   **Winemaking.**

—Jonathan Swinchatt, *The Winemaker's Dance*, 2004, p. 197

☺   ☹   😐   **Wine appreciation.**

—Maureen Petrosky, *The Wine Club*, 2005, p. 9

☺   ☹   😐   **Parisian etiquette.**

—Steve Fallon, *Paris*, 2004, p. 14

ROCKET SCIENCE?

Yes	No	Maybe
☺	☹	😐

Communications.
—Torie Clarke, *Lipstick on a Pig*, 2006, p. 171

Yes	No	Maybe
☺	☹	😐

Mnemonics.
—P. Waddington, *The Violent Workplace*, 2006, p. 20

Yes	No	Maybe
☺	☹	😐

Bilingualism.
—Alan Davies, *The Handbook of Applied Linguistics*, 2004, p. 462

ROCKET SCIENCE?

Yes   No   Maybe

🙂   ☹️   😐   **Aquaculture.**

—John Mosig, *Australian Fish Farmer*, 2004, p. v

🙂   ☹️   😐   **Trolling.**

—Tim Mercer, *Trolling: It's Not Rocket Science*, 1995

🙂   ☹️   😐   **Cookbook writing.**

—Dianne Jacon, *Will Write for Food*, 2005, p. 31

ROCKET SCIENCE?

Yes	No	Maybe	
☺	☹	😐	**Storytelling.**

—Annette Simmons, *The Story Factor*, 2006, p. 26

☺	☹	😐	**Backing up data.**

—Luke Welling, *PHP and MySQL Web Development*, 2003, p. 277

☺	☹	😐	**Eliminating office drama.**

—Debra Mandel, *Your Boss in Not Your Mother*, 2006, p. 19

ROCKET SCIENCE?

Yes   No   Maybe

☺   ☹   😐   **Acting in commercials.**

—Tom Logan, *Acting in the Million Dollar Minute, 2005*, p. 14

☺   ☹   😐   **Working with difficult children.**

—Dave Ziegler, *Achieving Success with Impossible Children*, 2005, p. 336

ROCKET SCIENCE?

Yes   No   Maybe

☺   ☹   😐   **Real estate investing.**

—Ralph Roberts, *Foreclosure Investing for Dummies*, 2007, p. 325

☺   ☹   😐   **Growing green lawns.**

—Jerry Baker, *Green Grass Magic*, 2004, p. 5

ROCKET SCIENCE?

Yes   No   Maybe

☺   ☹   😐   **Turning an abusive relationship into a compassionate one.**
—Steven Stosny, 2006, p. 207

☺   ☹   😐   **Basic self-defense.**
—Clois Williams, *Aircrew Security*, 2004, p. 385

ROCKET SCIENCE?

Yes   No   Maybe

☺   ☹   😐   **Basketball.**

—Dean Smith, *The Carolina Way*, 2004, p. 239

☺   ☹   😐   **Football.**

—Tiki Barber, *Tiki*, 2007, p. 193

☺   ☹   😐   **Self-coaching.**

—Talane Miedaner, *Coach Yourself to Success*, 2000, p. 74

ROCKET SCIENCE?

Yes   No   Maybe

☺   ☹   😐   **The fine and practiced art to preparing your cigar for smoking.**
—Tad Gage, *The Complete Idiot's Guide to Cigars*, 1997, p. 25

☺   ☹   😐   **Making a good humidor.**
—Fine Woodworking Magazine, *Building Small Projects*, 2004, p. 32

ROCKET SCIENCE?

Yes   No   Maybe

☺   ☹   😐   **Creating your life from the inside out.**

—Phillip McGraw, *Self Matters*, 2003, p. 204

☺   ☹   😐   **Evolution.**

—Greg Krukonis, *Evolution for Dummies*, 2008, p. 1

ROCKET SCIENCE?

Yes No Maybe

☺ ☹ 😐 **Five billion years of global change.**

—Denis Wood, *Five Billion Years of Global Change*, 2004, p. 156

☺ ☹ 😐 **Empirical science.**

—Keith Stanovich, *Progress in Understanding Reading*, 2000, p. 410

☺ ☹ 😐 **Knowledge elicitation.**

—R. Roy, *Industrial Knowledge Management*, 2001, p. 449

☺ ☹ 😐 **Authoring a doctoral thesis.**

—Patrick Dunleavy, *Authoring a PhD*, 2003, p. 68

ROCKET SCIENCE?

Yes	No	Maybe	
☺	☹	😐	**Looking at the conflicts in your life from a different perspective.** —Sandra Crowe, *Since Strangling Isn't an Option*, 1999, p. 10
☺	☹	😐	**Discovering your destiny.** —Janet Attwood, *The Passion Test*, 2007, p. 258
☺	☹	😐	**Rebuilding a resilient nation.** —Stephen Flynn, *The Edge of Disaster*, 2007, p. 98

ROCKET SCIENCE?

Yes No Maybe

☺ ☹ 😐 **Clapboard nailing.**

—Michael Litchfield, *Renovation*, 2005, p. 140

☺ ☹ 😐 **Benchmarking.**

—Brian Atkin, *Total Facilities Management*, 2005, p. 160

☺ ☹ 😐 **Building a wooden boat.**

—Greg Rossel, *Building Small Boats*, 1998, p. 1

☺ ☹ 😐 **Caliper overhaul.**

—Lee Klancher, *How to Customize Damn Near Anything*, 2003, p. 98

ROCKET SCIENCE?

Yes   No   Maybe

☺   ☹   😐   **Radiology.**

      —Richard Daffner, *Clinical Radiology*, 2007, p. 527

☺   ☹   😐   **Breaking good news.**

      —Robert Ramsey, *How to Say the Right Thing Every Time*, 2002, p. 71

ROCKET SCIENCE?

Yes   No   Maybe

☺   ☹   😐   **Chemistry.**
—Bonny Wolf, *Talking with My Mouth Full*, 2006, p. 58

☺   ☹   😐   **Biotechnology.**
—Ton Abate, *The Biotech Investor*, 2003, p. xii

☺   ☹   😐   **Epidemiology.**
—Elizabeth Pisani, *The Wisdom of Whores*, 2008, p. 315

ROCKET SCIENCE?

Yes   No   Maybe

☺   ☹   😐 **Bowling.**

        —Patricia Farrell, *How to Be Your Own Therapist*, 2004, p. 234

☺   ☹   😐 **Stopping activities that weaken you.**

        —Marcus Buckingham, *Go Put Your Strengths to Work*, 2007. p. 226

☺   ☹   😐 **Taking proper care of annuals.**

        —Steven Frowine, *Gardening Basics for Dummies*, 2007, p. 14

ROCKET SCIENCE?

Yes	No	Maybe
☺	☹	😐

Getting your radio tuner activated.

—E. Haley, *Over-the-Road Wireless for Dummies*, 2006, p. 305

☺	☹	😐

Singing.

—Phyllis Fulford, *The Complete Idiot's Guide to Singing*, 2003, p. xvii

ROCKET SCIENCE?

Yes   No   Maybe

☺ ☹ 😐 **Living a richer life with less stuff.**

—Peter Walsh, *It's All Too Much*, 2006, p. 56

☺ ☹ 😐 **Selling a house.**

—Eric Tyson, *House Selling for Dummies*, 2007, p. 91

☺ ☹ 😐 **Becoming invisible.**

—Barbara Delinsky, *An Accidental Woman*, 2002, p. 156

ROCKET SCIENCE?

Yes   No   Maybe

☺   ☹   😐   **The biology of muscle.**

        —Jeff Volek, *Men's Health TNT Diet*, 2007, p. 132

☺   ☹   😐   **Modeling.**

        —Olivia Goldsmith, *Fashionably Late*, 1995, p. 239

ROCKET SCIENCE?

Yes   No Maybe

☺   ☹   😐   **Retirement.**

         —Charles Grassley, *Living Longer, Retiring Earlier*, 2000, p. 32

☺   ☹   😐   **Harley-Davidson reassembly.**

         —Kenna Love, *101 Harley-Davidson Performance Projects*, 1998, p. 129

☺   ☹   😐   **Golf.**

         —Bobby Clampett, *The Impact Zone*, 2007, p. 113

ROCKET SCIENCE?

Yes   No   Maybe

☺ ☹ 😐 **Good hospitality.**

—*Pauline Frommer's Walt Disney World & Orlando*, 2007, p. 239

☺ ☹ 😐 **Avoiding junk carbs.**

—Gregg Avedon, *Men's Health Muscle Chow*, 2007, p. 9

☺ ☹ 😐 **Popping popcorn.**

—David Jones, *Candy Making for Dummies*, 2005, p. 156

ROCKET SCIENCE?

Yes	No	Maybe	
☺	☹	😐	**The secret to success.**

—Phil Rosenzweig, *The Halo Effect*, 2007, p. 86

Yes	No	Maybe	
☺	☹	😐	**Pleasuring a woman both in and out of bed.**

—Daylle Schwartz, *How to Please a Women in and Out of Bed*, 2005, p. 216

Yes	No	Maybe	
☺	☹	😐	**The scientific method.**

—Steven Milloy, *Junk Science Judo*, 2001, p. 42

Yes	No	Maybe	
☺	☹	😐	**Preventing sexually transmitted diseases.**

—David Clayton, *The Healthy Guide to Unhealthy Living*, 2005, p. 125

ROCKET SCIENCE?

Yes No Maybe

☺ ☹ 😐 **The science of air, scent, wind, and non-scents.**

—Kathy Etling, *The Art of Whitetail Deception*, 2002, p. 74

☺ ☹ 😐 **Aromatherapy.**

—Kim Morrison, *Like Chocolate for Women*, 2003, p. 16

ROCKET SCIENCE?

Yes No Maybe

☺ ☹ 😐 **Getting a man you like to do what you want.**

—Wendy Keller, *Secrets of Successful Negotiating for Women*, 2004, p. 208

☺ ☹ 😐 **Fixing a leaky washer.**

—Lee Jackson, *Redemption*, 2007, p. 29

☺ ☹ 😐 **Gaining two hours a day.**

—Vince Panella, *The 26-Hour Day*, 2001, p. 193

ROCKET SCIENCE?

Yes   No   Maybe

☺ ☹ 😐 **Opera.**

—Michael Walsh, *Who's Afraid of Opera?*, 1994, p. 213

☺ ☹ 😐 **Ushering.**

—James Grace, *The Best Man's Handbook*, 2004, p. 94

ROCKET SCIENCE?

Yes   No   Maybe

☺ ☹ 😐 **Getting kids to eat fruit.**

—Guy Cook, *The Discourse of Advertising*, 2001, p. 166

☺ ☹ 😐 **Plump, juicy raspberries in a graham cracker crust.**

—Elaine Balliet, *Blue*, 2007, p. 243

☺ ☹ 😐 **Liquid hand soap.**

—Harvard Business School, *Harvard Business Essentials*, 2005, p. 55

ROCKET SCIENCE?

Yes   No   Maybe

☺   ☹   😐   **Fundraising.**

—Ken Burnett, *The Zen of Fundraising*, 2006, p. 112

☺   ☹   😐   **Casino gambling.**

—Andrew Brisman, *Mensa Guide to Casino Gambling*, 2004, p. 101

☺   ☹   😐   **Coming up with a salvage plan.**

—Alan Brunacini, *Fire Command*, 2002, p. 416

ROCKET SCIENCE?

Yes   No   Maybe

☺ ☹ 😐 **Mixing powder with water.**

—Paige Hobey, *The Working Gal's Guide to Babyville*, 2006, p. 85

☺ ☹ 😐 **Watercoloring.**

—Jack Reid, *Watercolor Basics*, 1998, p. 47

ROCKET SCIENCE?

Yes   No   Maybe

☺   ☹   😐   **Following in Mom's footsteps.**

        —Yvonne Collins, *Introducing Vivien Leigh Reid*, 2005, p. 57

☺   ☹   😐   **Driving from point A to point B on an interstate.**

        —Bob Miller, *RV*, 2007, p. 63

ROCKET SCIENCE?

Yes   No   Maybe

☺   ☹   😐   **Producing a reality show.**

                —Bill Bryan, *Keep it Real*, 2007, p. 26

☺   ☹   😐   **Infrared photography.**

                —Joe France, *Complete Guide to Digital Infrared Photography*, 2006, p. 24

ROCKET SCIENCE?

Yes   No   Maybe

☺   ☹   😐   **Choosing what to eat.**

—Kate Cook, *Boost Your Whole Health*, 2007, p. 77

☺   ☹   😐   **Opening a boiled crab.**

—Andrew Jaeger, *New Orleans Seafood Cookbook*, 1999, p. 74

☺   ☹   😐   **Making a salad.**

—Kenneth Goldberg, *The Men's Health Longevity Program*, 2001, p. 28

ROCKET SCIENCE?

Yes   No   Maybe

☺   ☹   😐   **Social research.**

—Daniel Chambliss, *Making Sense of the Social World*, 2006, p. 18

☺   ☹   😐   **Human behavior.**

—University of California, *American Short Fiction*, 1981, p. 247

☺   ☹   😐   **Infidelity.**

—David Barash, *The Myth of Monogamy*, 2002, p. 16

ROCKET SCIENCE?

Yes   No   Maybe

☺   ☹   😐   **Yoga.**

—Steve Ross, *Happy Yoga*, 2003, p. 5

☺   ☹   😐   **Meditation.**

—Loretta LaRoche, *Kick Up Your Heels*, 2008, p. 108

☺   ☹   😐   **Dharma.**

—David Guy, *Jake Fades*, 2007, p. 25

☺   ☹   😐   **Spiritual growth.**

—Alex Gee, *Jesus and the Hip-Hop Prophets*, 2003, p. 107

ROCKET SCIENCE?

Yes No Maybe

☺ ☹ 😐 **Hypnosis.**

—Jonathan Chase, *Don't Look in His Eyes!*, 2007, p. 5

☺ ☹ 😐 **Reshaping attitudes.**

—David Lieberman, *How to Change Anybody*, 2006, p. 114

ROCKET SCIENCE?

Yes   No   Maybe

☺   ☹   😐   **Warming up.**

—Sam Murphy, *Run for Life*, 2004, p. 49

☺   ☹   😐   **Changing a banjo string.**

—Bill Evans, *Banjo for Dummies*, 2007, p. 252

ROCKET SCIENCE?

Yes   No   Maybe

☺   ☹   😐   **Wiring networks.**

—Bill Ferguson, *Network+ Fast Pass*, 2005, p. 120

☺   ☹   😐   **Wireless networks.**

—Jim Aspinwall, *Installing, Troubleshooting, and Repairing Wireless Networks*, 2003

ROCKET SCIENCE?

Yes   No   Maybe

☺   ☹   😐   **Identifying patterns in blood sugar readings.**

—Virginia Valentine, *Diabetes Type 2*, 1998, p. 44

☺   ☹   😐   **Beating diabetes.**

—David Nathan, *Beating Diabetes,* 2005, p. 202

ROCKET SCIENCE?

Yes No Maybe

☺ ☹ 😐 **Planting trees.**

—National Association of Regulatory Utility Commissioners, 1993, p. 262

☺ ☹ 😐 **Iguanas.**

—Melissa Kaplan, *Iguanas for Dummies*, 2000, p. 1

☺ ☹ 😐 **Bark-peeling.**

—Eugene Buchanan, *Brothers on the Bashkaus*, 2007, p. 79

ROCKET SCIENCE?

Yes   No   Maybe

☺   ☹   😐   **Operating a gas-powered auger.**

—Noel Vick, *Fishing on Ice*, 1999, p. 196

☺   ☹   😐   **Drilling shelf holes.**

—Joe Hurst-Wajszczuk, *Furniture You Can Build*, 2006, p. 118

ROCKET SCIENCE?

Yes   No   Maybe

☺   ☹   😐   **Deconstructing special, expert knowledge.**

—Barry Duncan, *The Heroic Client*, 2004, p. 197

☺   ☹   😐   **Learning the rules and conventions of language.**

—Stephen Wilbers, *Keys to Great Writing*, 2007, p. 240

ROCKET SCIENCE?

☺   ☹   😐   **Cleaning a knife properly.**

—Sur La Table, *Knifes Books Love*, 2008, p. 77

☺   ☹   😐   **Cutting cacti.**

—Shirley-Ann Bell, *Success with Cacti and Other Succulents*, 2006, p. 93

ROCKET SCIENCE?

Yes   No   Maybe

😊   😞   😐   **Ergonomics.**

—Waldemar Karwowski, *The Occupational Ergonomics Handbook*, 1999, p. 76

😊   😞   😐   **Feltmaking.**

—Robyn Steel-Strickland, *Felt*, 2006, p. 1

ROCKET SCIENCE?

Yes   No   Maybe

☺   ☹   😐   **Using a business card.**
       —Bob Popyk, *Here's My Card*, 2000, p. 137

☺   ☹   😐   **Testifying as an expert.**
       —Derrick Watkins, *Gang Investigations*, 2006, p. 96

ROCKET SCIENCE?

Yes   No   Maybe

☺   ☹   😐   **A tidy living room.**

—Jane Bullivant, *Dear Lord, I Feel Like a Whale*, 2005, p. 96

☺   ☹   😐   **Entropy.**

—Stephen Haines, *Enterprise-wide Change*, 2004, p. 253

ROCKET SCIENCE?

Yes   No   Maybe

☺   ☹   😐   **Training individuals not associated with rocket science.**

—Judith Harrington, *The Everything Start Your Own Business Book*, 2006, p. 275

☺   ☹   😐   **Art appreciation.**

—Julian Stallabrass, *Art Incorporated*, 2004, p. 170

ANSWER KEY

All answers are "Not rocket science" except:

  The race for space.

—Betsy Kuhn, *The Race for Space*, 2007, p. 24

  Launching, flying and returning the Space Shuttle.

—Fergus O'Connell, *Simply Brilliant*, 2004, p. 5